Timothy Richards Lewis

On a Haematozoon Inhabiting Human Blood

Timothy Richards Lewis

On a Haematozoon Inhabiting Human Blood

ISBN/EAN: 9783337366087

Printed in Europe, USA, Canada, Australia, Japan

Cover: Foto ©berggeist007 / pixelio.de

More available books at **www.hansebooks.com**

ON A

HÆMATOZOON

INHABITING HUMAN BLOOD:

ITS RELATION TO

CHYLURIA AND OTHER DISEASES.

BY

TIMOTHY RICHARDS LEWIS, M.B.,

ASSISTANT-SURGEON, H.M. BRITISH FORCES;

(ON SPECIAL DUTY)

ATTACHED TO THE SANITARY COMMISSIONER WITH THE GOVERNMENT OF INDIA.

CALCUTTA:

OFFICE OF THE SUPERINTENDENT OF GOVERNMENT PRINTING.

1872.

CONTENTS.

Wood-Engravings.

HÆMATOZOON

INHABITING HUMAN BLOOD:

ITS RELATION TO

CHYLURIA AND OTHER DISEASES.*

FOR many generations writers on medical subjects have

The belief in the existence of Human Hæmatozoa long entertained.

maintained that the human blood during certain diseased conditions is invaded by parasites. The opinion most in favour has been, that these in all probability were in the form of worms; but, so far as I have been able to ascertain, it has never yet been satisfactorily demonstrated that this condition really existed.

That certain limited areas of the circulatory tract

The Distomata hitherto discovered are too large to pass through the capillaries;

may become invaded by Entozoa has long been known : the portal vein, and the vessels in more or less direct relation with the intestinal canal, are the channels which have usually been thus affected; but the parasites found in these situations, such as the Distoma hæmatobium,

---

* Forming an Appendix to the Eighth Annual Report of the Sanitary Commissioner with the Government of India.

discovered by Bilharz in 1851, and a few other imperfectly described distomata, are far too large to pass through any but comparatively capacious blood vessels. The instances on record in which they have been found in vessels beyond these limits are few, and evidently accidental occurrences. None of these, therefore, can, I think, be justly described as 'Hæmatozoa' in the strict sense of the term.

The same remarks apply, with only very slight modifications, to the presence of Echinococci in the blood-vessels, a few young specimens of which, have, on rare occasions, been discovered (by Klencke and others) in the general circulation, but then only in vessels of considerable calibre.

So likewise are Echinococci.

It has also been inferred that the progeny of some Entozoa must be carried by the blood-current, as otherwise they could have not reached their destination so rapidly in the various distant parts of the body in which they have been found. That the Trichina spiralis, for example, during its earlier migrations, may be conveyed in this way, is, although strongly denied, I think not improbable. As their presence in the blood has not, however, been recognized, either in man or in animals, their sojourn in such channels, must, at all events, be of very short duration.

Probability of some parasites having reached the tissues in which they are found, by means of the blood-vessels.

But that a condition should exist in which human blood should be infested by living active worms in either an embryo or mature state, to the extent hereafter to be described,

The discovery of microscopic worms in great numbers in human blood.

had, I presume, scarcely been surmised—a condition in which they are persistently so ubiquitous as to be obtained day after day in numbers, by simply pricking any portion of the body, even to the tips of the fingers and toes of both hands and both feet of one and the same person with a finely pointed needle. On one occasion six excellent specimens were obtained in a single drop of blood, by merely pricking the lobule of the ear.

Towards the beginning of July of the present year, whilst examining the blood of a native suffering from diarrhœa, a patient at the Medical College Hospital under Dr. Chuckerbutty's care, I observed nine minute Nematoid worms in a state of great activity, on a single slide. On drawing the attention of my colleague, Dr. Douglas Cunningham, to the preparation, he fully coincided in my opinion that they were precisely the same kind as those observed by me more than two years previously (in March 1870), as being constantly present in Chylous urine.

Date of their discovery in the blood, and of their discovery in the urine.

In a report on the microscopic characters of choleraic dejecta published at the time, both separately and also in the form of an Appendix to the Sixth Report of the Sanitary Commissioner with the Government of India, I had occasion to allude to this condition of the urine in connection with some cells observed in it, which closely corresponded in appearance with cells constantly present in choleraic discharges, and the opportunity was taken of drawing attention to the Entozoon, which was at the same time figured and described.

A synopsis of the first case published.

For the sake of convenience it may be well to refer to this case again. The patient was a deaf and emaciated East Indian, about 25 years of age, under the care of Mr. R. T. Lyons at the General Hospital, and was kept under observation for a period of two months, during which time his urine continued to present a white, milky appearance, and yellowish-white coagula rapidly formed in the vessel into which it had been voided. When a small portion of the gelatinised substance was teased with needles on a slide, and placed under the microscope, delicate filaments were seen, partly hidden by the fibro-albuminous matter in which they were embedded, and which I at first considered to be scattered filaments of a growing fungus. After being watched for some time, however, they were seen to coil and uncoil themselves, so that all doubt as to their nature was at an end. I had opportunities of showing them on various occasions to several persons; and having perfectly satisfied myself that their occurrence was not accidental, nor yet the result of subsequent development in the urine, after the manner of the Anguillulæ which are so well known to develop in vinegar or starch-paste, I did not hesitate to draw attention to them as being the probable cause of the obscure disease known as " Chyluria."*

From this period I have paid considerable attention to the subject, and I desire to express the obligations I am under to Dr. Ewart, the Surgeon in charge of the Presi-

How and where subsequent cases were obtained.

---

* Subsequent observations in connection with this case will be found referred to further on—p. 46.

dency General Hospital; to Dr. D. B. Smith, the Officiating Principal of the Medical College; to Dr. Edmonston Charles, Professor of Midwifery at the same College, and to Dr. McConnell, the Professor of Pathology, as well as to several others, for the opportunities afforded me for the study of this and of allied conditions of the urine.

A slide containing one or more specimens of this Nematode having been forwarded to Professor Parkes, at Netley, he very kindly showed it to Mr. Busk, whose

Probable group of Nematodes to which the Hæmatozoon belongs.

extensive knowledge in this department of science is well known, and the opinion was expressed by him, that, so far as could be judged from the form and size alone, the worm seemed to belong to the *Filaridæ*.

At this time it was not known to exist in the blood, nor had its minute anatomy been accurately ascertained; however, I do not anticipate that the information acquired since that time would materially alter Mr. Busk's opinion, so that perhaps the name already applied to the Hæmatozoon in the columns of the 'Lancet,' *Filaria Sanguinis hominis*, may provisionally be adopted.

I am indebted for the greater number of the specimens of Chylous urine which I have examined to Dr. Charles, who and Dr. W. J. Palmer, were, I believe, the first to verify the observations which I had published, both having had cases of the disease about the same time towards the end of 1870 or beginning of 1871. The fact of Dr. Charles being in charge of the midwifery wards of the College Hospital,

Chyluria appears to be more prevalent among women than among men. has apparently conduced to his being able to aid me so materially, as, strange to say, the patients suffering from Chyluria have, for the most part, been women; in the last case brought to my notice by him, this condition was observed, for the first time, four days after podalic version had been performed.

I have now observed the urine in this condition, *Number of cases of Chyluria observed.* associated with more or less marked hæmaturia, in from fifteen to twenty patients, several samples having been obtained from nearly all of them; *these microscopic Filariæ have been present on every occasion.* Of the persons thus affected, five were ascertained to be of pure European parentage, but three of them were born in this country; the remainder were either East Indians or Natives, in about equal proportion.

I regret that I lost the opportunity of fully ascertain- *Particulars concerning the patient in whom the Hæmatozoa were first discovered.* ing the previous history of the case in which the Hæmatozoa were first detected. Having satisfied myself of the identity of the worms previously observed in the urine and now in the blood, by careful comparison of their form, structure and measurement, I returned on the following morning to the Medical College for this purpose, but to my great disappointment found that the man had been discharged, at his own request, an hour before my arrival. He had, it appears, suffered from diarrhœa for about a fortnight, which had become greatly aggravated a few days before his admission into hospital; but nothing further could be learnt of the state of his health beyond that he had complained of deafness, especially of one ear.*

He had left no address, except that he was a blacksmith living in a large bazaar in the neighbourhood; but as some three or four thousand persons are crowded into this bazaar, a great proportion of whom are smiths in some form or another, those acquainted with the intricate geography of such places in the East will not be surprised to learn that I spent a whole afternoon searching for him in vain. I then enlisted the friendly aid of the Police, but this also proved fruitless.

A few days after this occurrence, Dr. D. B. Smith informed me that there was a native

A second case of Filariæ in the blood, associated with Chyluria.

woman in one of his wards suffering from hæmaturia, combined with a chylous condition of the urine, and very kindly forwarded a specimen of it on the following morning; this urine, as usual under such circumstances, contained the worms in abundance.

I saw the woman on the evening of the same day, and learnt that the complaint from

First and second attack.

which she was suffering had first made its appearance during the third month of her last pregnancy, but that it had apparently passed off in about five or six weeks. After the birth of this child, which

---

* One of the patients brought to my notice, who had suffered from Chyluria for several months, had been in hospital for another complaint, and had actually left the hospital without having mentioned a word about the condition of his urine. He stated afterwards that he did not like to do so, as it was no great inconvenience to him, and he imagined it was the temporary result of an 'indiscretion.'

was now five months old, the disease came on again : she was unable to suckle her infant, the lacteal secretion being altogether absent.*

On pricking her finger with a needle, and distributing a drop of blood over several slides, I found that the Filariæ were present in it also.

She remained under observation for a period of about two months, but there was no mark-

Other symptoms, and progress of disease.

ed change in her general condition; her face, however, became swollen on one or two occasions, and appeared puffy, as also did the upper and lower extremities. The urine slightly improved in appearance, and the numbers of the Filariæ in it as well as in the blood diminished; in the latter especially, at all events, the numbers obtainable by pricking the fingers or toes certainly decreased; and eventually, out of half a dozen or more slides, not more than one or two Hæmatozoa could be detected : on a few occasions several slides were examined without any being found.

The patient, however, could not be persuaded to remain longer in hospital: indeed, all patients thus affected soon get tired of being treated for their complaint, as there is seldom any great suffering which the patient can directly connect with this condition, and often no other very well marked symptom, beyond general debility.

---

* A case is recorded by Drs. Mayer and Pearse as having occurred in the Madras Presidency, in which a young East Indian woman had suffered after two pregnancies in this manner; she continued to suckle her children uniuterruptedly, but on being advised on the last occasion to discontinue doing so, the urine returned to its natural appearance. *Brit. and For. Med.-Chir. Review,* Vol. IX, 1852, page 511.

The most remarkable case which has come under my notice, in which the blood was affected in this manner, was that of a patient in one of Dr. Ewart's wards, whom he kindly placed at my disposal for observation and treatment. The man was an East Indian (with more of the habits of the native than of the European), about 22 years of age; he had been for about five years employed as cook on board one of the light-ships lying at the entrance of the Hooghly, spending only about three months of the year with his family on shore.

A third case of Filaria in the blood, associated with Chyluria.

The prominent symptoms in this case were, extreme and persistent milkiness of the urine, which coagulated with great rapidity after being voided. On being heated the smell given off at first corresponded exactly with that of warm milk, but when the heat was continued for some time, was gradually replaced by the ordinary smell of urine. This condition came on *suddenly* about two months previous to his admission into the General Hospital.

He dates his illness, however, as having commenced about a year before, for his sight then became affected, and there was 'inflammation' of both eyes, together with a copious discharge of fluid from them. These symptoms have persisted, although he thinks that they have somewhat subsided since the change occurred in his urine. He has well marked 'granular lids,' the mucous membrane of both the upper and lower lids are red and swollen, and the sclerotic conjunctiva injected, the vessels being large and tortuous; there is also considerable opacity

Other prominent symptoms.

of the cornea. He presents a somewhat emaciated appearance, although his appetite had always continued good, and certainly since his admission into hospital the man has gladly availed himself of the most liberal scale of diet allowed.

This is not surprising when the amount of fibro-albuminous matter which is con-

Extreme extent of infection.

stantly being drained from his system, as evidenced by the state of the urinary secretion, is taken into consideration ; but when to this is added the fact that no matter at what portion of his body the circulation is tapped with the point of a needle, numerous active, well-developed Hæmatozoa are invariably obtained : on one occasion I obtained as many as twelve of these creatures on a single slide. As the drop of blood from which this slide had been prepared sufficed for the preparation of two or three other slides (which, however, between them did not contain more than half a dozen Filariæ), the number infesting his whole body may be imagined.

A rough calculation may very readily be made ; the weight of the man is 100℔; if the

Approximate number of Hæmatozoa present in the body.

amount of blood be taken as being on the average 'not less than one-tenth of the weight of the body' (Huxley), and it is assumed that each drop, or grain rather, contains, say, two Hæmatozoa, it would be but a reasonable estimate to set down the number as 140,000 ! It must, however, be borne in mind, that the Hæmatozoa may be more or less localised to the capillaries and smaller vessels, which would materially modify this estimate, still I know of no fact which warrants any such assumption.

The urine also contained numerous Filariæ, but they were by no means so plentiful in this fluid as the con- The number in the urine. dition of the blood might have led one to expect. I have seen them far more plentiful in the urinary secretion of a person whose blood did not appear to be infected to anything like the extent to which this man's had been.

On several occasions I attempted, but in vain, to detect the Filariæ in the copious slightly milky secretion constantly welling out of the corner Chylous (?) discharge from the lachrymal and adjoining glands. of his eyes, and which in a slight degree appeared to curdle. I feel convinced, however, that could a sufficient quantity of this secretion be accumulated they would be discovered.* Microscopically the discharge consisted of clear fluid with numerous granular cells, precisely as observed in the urine of persons suffering from Chyluria. The reaction of the fluid was slightly alkaline.

Before alluding more minutely to the appearances presented by the Hæmatozoon, I will A fourth case of Hæmatozoa associated with Chyluria. refer to one other case for the opportunity of observing which I am again indebted to the Principal of the Medical College, under

---

* Since this portion was in type the inference above made has proved to have been correct, as I have obtained a specimen of the Filariæ in the midst of a shred of flocculent matter, which had been transferred from the inner surface of one of the lower eye-lids on to a glass slide for examination.

Its breadth was $\frac{1}{1800}$ of an inch and its length $\frac{1}{35}$", the relative proportion between the latter and the former being, therefore, as 1 to 52.

whose care the patient was. She was the wife of a police sergeant, 30 to 35 years of age, born in this country,

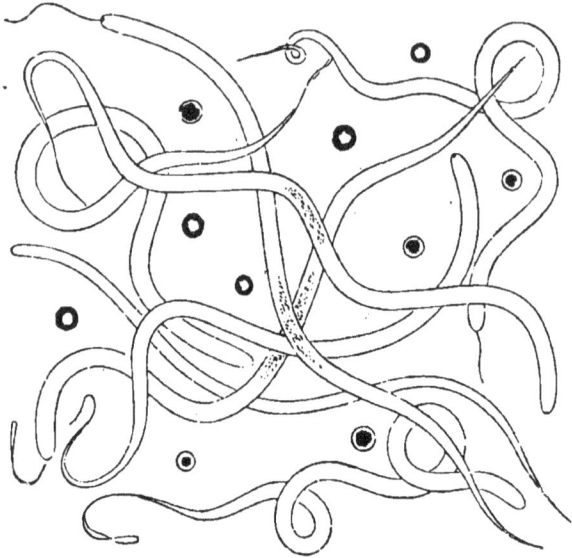

Living Hæmatozon observed in a single preparation of blood obtained by pricking (with a needle) the finger of a European woman suffering from Chyluria.

A few red-blood corpuscles have been introduced to show the relative size of the Filariæ.

but of pure European parentage. Towards the end of July last she was admitted into hospital suffering from a chylous condition of the urine, with frequently recurring attacks of hæmaturia.

The disease made its appearance two months after First attack.   child-birth, 16 years ago, when she was living a few miles from Calcutta—at Hooghly. It continued for six months, and, in her opinion, was cured by taking an infusion of the seed of an aromatic plant used by the natives for flavouring curries called "*kahlajeera*," a species of Nigella (sativa?).

In the following year she was again confined, but
Second attack. the symptoms did not return; in
1859, whilst residing at Rajshayc, the disease reappeared.
She was then neither pregnant nor nursing. In three
and a half months the symptoms subsided, the above-
named native remedy having been administered as before.
Since this period she had given birth to two more
Third attack. children, the last child having been
born in 1864; but no symptoms of her complaint had
appeared until within a few days of her admission into
Dr. Smith's ward, when they came on suddenly after a
lapse of eight years. During the first three weeks of
her stay in hospital no marked alteration in her condition was observed, neither for better nor for worse—
Hæmatozoa were persistently present in her blood, no
matter from what portion of her body the fluid was
obtained; they were also present in the urine.

Dr. Smith tried muriate of iron, gallic acid, as well
Fatal termination. as numerous other remedial agents,
mineral and vegetable, not omitting the " kahlajeera,"
in which she had much faith, but none of them seemed
to produce the slightest effect. The proportion of blood
in the urine increased, painful diarrhœa set in, with
rapid emaciation, and she died about six weeks after her
admission.

It was with considerable difficulty that permission
was obtained to make a *post-mortem*
Post-mortem examin-
ation. examination, which had moreover to
be so hurriedly performed that Dr. McConnell, the
Professor of Pathology, was unable to give me notice;
but he has most kindly placed at my disposal the careful

notes which he made of the appearances presented by the various viscera (fourteen hours after death), a summary of which is here given.

The brain was soft and somewhat anæmic; other-

Condition of the brain and of the thoracic viscera.

wise there was nothing special to be observed in its structure, nor in its ventricles. The right side of the heart contained small semi-decolorised clots, as also did the left auricle, but the ventricle was empty. There was some thickening of the mitral valve, and slight, irregular thickening of the lining membrane of the aorta—further than this there seemed to be nothing abnormal. The mucous surface of the trachea and bronchi appeared to be healthy. Scattered throughout the whole of the right lung were numerous specks of what appeared to be softening tubercle, each about the size of a pea; in addition to which two circumscribed cavities, one of the size of a hen's egg, the other about half that size, were found in the substance of the middle lobe; each cavity was lined by a distinct "pyogenic membrane" and contained thick muco-pus. The left lung contained a few small, irregularly distributed nodules of softening tuberculous matter, and one cavity, the size of a pigeon's egg, filled with muco-pus. The weight of the right lung was 5 ounces and 6 drachms, and that of the left was 8 ounces and 4 drachms.

The mucous membrane of the stomach was of a

Condition of the intestine and other abdominal viscera.

bright pink colour, not altered in consistence, whereas the mucous surface of the duodenum presented a mammilated and congested appearance. The jejunum and ileum in the upper half were healthy, but in the lower half of the latter the Peyer's patches were prominent, and the surface covered with minute ulcers—the glandules infiltrated with a yellowish-white, soft, tubercular-like substance; the edges of the ulcers thickened and containing similar yellowish-white granular matter. The lining membrane of the cœcum and ascending colon was of a bright pink colour, and exhibited five or six circular ulcers about the size of half a pea, with raised and opaque white edges. The entire contents of the intestines consisted only of about a couple of ounces of a pea-soup-like fluid. The mesenteric glands were unaffected.

The liver was soft and fatty, otherwise normal in appearance; no re-action with iodine. The gall-bladder was almost empty, containing only a little thin ochre-coloured bile. The spleen seemed to be healthy, as did the uterus and ovaries; the former was small and unimpregnated. The kidneys presented nothing abnormal to the naked eye; the right and left weighed respectively 3 ounces 4 drachms, and 3 ounces 6 drachms: these, together with the supra-renal capsules, were placed in spirit and kindly forwarded by Dr. McConnell for my examination, the result of which will be referred to further on (page 33).

As in other cases of Chyluria on record, so in this, not the remotest clue is afforded as to the nature, or as to the cause of the disease by the *post-mortem* appearances visible to the naked eye; nor is there any sufficiently-marked lesion to account for the condition of the urine during life, nor for the rapid manner in which the patient ultimately succumbed.

In order to detect the Hæmatozoon during life, the

Method adopted for the detection of Hæmatozoa. method adopted by me is as follows :—A. piece of narrow tape is coiled tightly round the end of one of the fingers or toes, so as to produce a little temporary local congestion of the part, but not to such an extent as to cause the slightest pain; and with a clean, finely-pointed needle the finger is gently pricked—half-a-dozen slides and covering glasses having been previously prepared. The drop of blood thus obtained will suffice for several slides, but I find it a good plan to squeeze out only a very small drop, and transfer it altogether to one slide by drawing the edge of the covering glass along the tip of the finger so as to *scrape* off the 'droplet.' The glass is then carefully pressed on to the slide by a gliding motion, in order to produce as thin a layer of fluid as possible, and to ensure that all the fluid removed is retained beneath the cover, because there is a tendency on the part of the fluid to carry the Hæmatozoa towards the edge of the slide, just as is observed to take place in examinations of the urine for 'casts' of the kidney-tubules.

The slides are now to be carefully and systematically examined ; a lateral and horizontal stage-movement being very useful for this purpose, as it enables us to make sure that every part of the preparation has been scrutinized.

It must not be supposed that the Filariæ are to be

Hæmatozoa even when present not always readily detected. detected by taking a mere peep through the microscope ; sometimes, certainly, I have observed one, or two even, in the first field examined; but this is by no means usual, and their detection often demands considerable patience. Each slide will require about a quarter of an hour before it can be satisfactorily explored ; any one who imagines that they can be detected with the same ease as a white-blood corpuscle had better not make the attempt.

Several slides may have to be examined, and it may be necessary to make a fresh puncture, for I have found the Hœmatozoa to be absent in several slides obtained from one finger, but present in all the slides obtained from another at the same time; whereas on making a fresh puncture in the finger where none had been found at first, it was ascertained that they were equally numerous in both. This is possibly accounted for by the little orifice made having become plugged by fat, &c., so that the blood squeezed through had to some extent been filtered, for although this microscopic Filaria can pass through any channel permeable to a red-blood corpuscle, still, when it is considered that the length of the former is nearly fifty times its diameter, the wonder is that they are not more completely prevented from escaping through so fine an orifice even when perfectly patent.

A possible cause of this difficulty.

The search should not be undertaken with too high a magnifying power, but it should be sufficiently high to define the outline of a red-blood corpuscle quite distinctly. I have found that a good two-thirds of an inch objective answers the purpose of a *searcher* admirably; it embraces a tolerably large area, so that the preparation can be examined in a comparatively short time; but care

The magnifying powers advisable to employ.

should be taken to keep the fine-adjustment screw constantly moving, so as to examine the deeper as well as the superficial layer of fluid in each field as it passes under observation. Should anything unusual be observed, the low power must be replaced by a $\frac{1}{4}''$ or, better, a $\frac{1}{8}''$ objective.*

In order to keep the active Hæmatozoon under observation for some hours, a camel-hair pencil dipped in a solution of Canada-balsam or mastic in chloroform, should be passed along the edges of the covering glass, so as to prevent evaporation, and the formation of air spaces in the preparation.

The appearance presented by the Hæmatozoon when first seen among the blood corpuscles, in the living state, will not readily be forgotten and cannot possibly be mistaken for anything else. The remark made by a young Bengalee student on my requesting him to look into the microscope and tell me what he saw—"He is an incompletely developed *snake*, evidently very young, though very active"—so aptly describes the object as thus witnessed, that I feel sure that any one seeing the

The general aspect of the Hæmatozoon during life.

---

* It may seem superfluous to draw attention to the similarity which exists between some vegetable fibres and some of the microscopic ' Filariæ,' when the latter are not alive; nevertheless a very good objective is frequently required to distinguish them with certainty, as any one may prove for himself by subjecting the torn fluffy edges of a piece of blotting-paper to microscopic examination. Filaria-like fibrils will frequently be found.

A mistake of this kind is referred to by Leuckart as having occurred quite recently. A paper was published announcing the discovery of a filaria-like nematode in the intestines, blood, and tissues of a patient, which was expected to prove as dangerous to life as the Trichina spiralis. These parasites subsequently proved to be nothing more than vegetable hairs! ' Die Menschlichen Parasiten,' Vol. II, part 1, p. 151.

Hæmatozoon alive will not fail to be struck with the accuracy of the quaint reply.

During the first few hours after removal from the body the Filaria is in constant motion, coiling and uncoiling itself unceasingly, lashing the blood corpuscles about in all directions, and insinuating itself between them. It is not at rest for a single moment, and yet on the slide it appears to make but little progress, as it may frequently be watched for an hour in the same field without once giving occasion to shift the stage of the microscope. No sooner has it insinuated its 'head' amongst a group of corpuscles than it is retracted, and probably the next instant the 'tail' will be darted forth and retracted in a similar way.

One moment it may appear to possess a long 'tail' —a fourth or more of its entire

Anatomical details visible under a ¼″ objective.

length, which follows it through the fluid like a string, whereas the very next moment not a trace of the 'tail' can be seen, even with the highest powers. The same phenomena can be observed to take place at the thicker, cephalic end, but with more difficulty. As usually seen, this presents a blunt or slightly tapering termination, but every now and then a fine point like a fang appears as if darted straight forward out of its substance ; the next instant the creature may jerk its 'head' on one side, and the 'fang' becomes bent and drawn after it like a ribbon (Fig. I, page 14).

As seen with a ¼″ objective, these Hæmatozoa can scarcely be said to present a granular aspect. When only recently withdrawn from the body, they look smooth and almost translucent ; the larger specimens, however, frequently present an aggregation of granules towards the junction of the middle with the lower half, as may be seen represented in a few of the specimens delineated (Fig. I). Occasionally also a bright clear spot is observed at the thicker extremity extremely suggestive of an oral aperture; this likewise is represented in some of the figures in the wood-cuts.

They will continue thus active under a covering Duration of their glass, hermetically sealed, for from 6 activity. to 24 or 30 hours, and if a drop of blood be suspended from the centre of a covering glass and fixed to a ring of wax, thus forming a closed cell, the Hæmatozoa may live for three days, perhaps longer, but this is the longest period during which I have known them to retain their activity.

It must not be inferred that the group of Hæmatozoa depicted in this wood-cut (Fig. 1) represents *one field* of the microscope, but only that the particular specimens were observed on a single slide. The same remark applies to the second group depicted; except, that two of the figures in it represent Filariæ found in other preparations, obtained from the same individual.

In the later periods of their existence the movements of the Filariæ become much slower, and the plasma of their bodies more and more granular until eventually all signs of activity disappear, and they are seen stretched or slightly curved in the field of the microscope, having lost the snake rather than worm-like appearance, which, from their tenuity and incessant coiling or wriggling movements, they had presented during life (Fig. II).

If a little spirit, or other preservative agent, such as corrosive sublimate or carbolic acid and glycerine be not added, their outline among the blood corpuscles will become indistinct, and they will degenerate into mere shrivelled strings of a granular appearance, no longer recognisable as Filariæ.

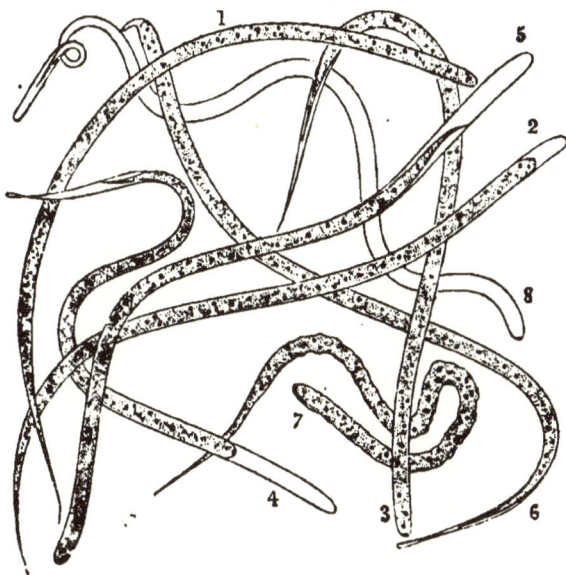

FIG. II.   × 300
HÆMATOZOA AS IN FIG. I, AFTER DEATH, BY THE ADDITION OF PRESERVATIVE MEDIA.

Nos. 1 to 6.   Preserved in weak spirit.
               (When the Filariæ are observed in slightly decomposed urine they
               present the appearances here shown also).
No. 7.   Puckered condition, produced by the first addition of pure glycerine.
 „ 8.   Killed by exposure to the fumes of Osmic acid.

Some of the various aspects presented by them after death are delineated in the second wood-cut.   In No. 1 (Fig. II) the Hæmatozoon presents a granular appearance throughout its entire length; but in No. 2, a hyaline membrane is seen to extend beyond the head extremity, and in

Variety of appearances presented by them after death.

No. 3 this transparent membrane appears as if it were a continuation of the tail; whereas in No. 4 it extends beyond them both. In No. 5 the membrane appears as if slightly wider at the 'tail' end, but is absent at the opposite extremity, and in No. 6 the membrane is bent in the form of a hook. In No. 7 it is seen puckered, on account of the addition of a thick fluid. The meaning of all the different appearances presented by these Filariæ, obtained from the same patient, will become evident on perusal of a succeeding paragraph.

One of the Hæmatozoa in this wood-cut (No. 8) is seen to have preserved the appearance presented during life, it having been instantaneously killed by holding the slide over the fumes of osmic acid—by far the best method I know for preserving the specimens. The blood should be quickly but evenly spread over the covering glass, forming as thin a layer as possible; the cover is then to be quickly inverted (before coagulation sets in) over the mouth of a phial containing a 2 per cent. aqueous solution of this chemical. When the preparation has turned somewhat brown, remove it and place it on a slide, previously charged with a drop of a saturated solution of acetate of potash or soda, when it is ready for mounting, and will keep, I believe, for an indefinite period.

The results of exposing them to the action of Osmic acid.

To account for the various appearances presented by these Hæmatozoa before and after death, which have been just described and figured, may possibly puzzle others as it certainly puzzled me for over two years,

The minute anatomy of the Hæmatozoon as observed under a ¼″ or ½″ immersion objective.

although I was constantly in the habit of examining specimens; but, until their existence in the blood had been discovered, by far the greater number of them had been dead, or nearly so, before they came under observation. Having observed that the appearance usually presented by the Hæmatozoa, when recently withdrawn from the circulation, differed considerably from what was observed after or shortly before death, it was determined to watch these changes from beginning to end, and to note them as they occurred. With this object in view, a specimen which appeared to be well developed, was selected out of several found on a quite recently prepared slide, a carefully corrected immersion $\frac{1}{8}''$ object-glass was employed, and the examination continued for eight consecutive hours.

At first the movements of the Hæmatozoon were
The existence of transverse striæ demonstrated. so rapid that little could be detected in addition to what had been quite as distinctly seen with $\frac{1}{4}''$ glass, except that in certain positions assumed by the worm, and in certain lights, extremely fine transverse striæ were observed quite distinctly. The existence of these striæ had, on several occasions, been more than suspected under the lower power ($\frac{1}{4}''$), but they could not be satisfactorily demonstrated. No attempt has been made to represent those fine markings in the wood-cuts as seen by such a, comparatively, low power as this, for it would only tend to mislead; to cut lines in wood only $\frac{1}{25000}$ of an inch apart (which is about the distance between the markings), when simply magnified 300 diameters, would be impossible; and even in the engraving which represents the

object as magnified by twice this power, the distinctness
of these markings is considerably exaggerated (Fig III).

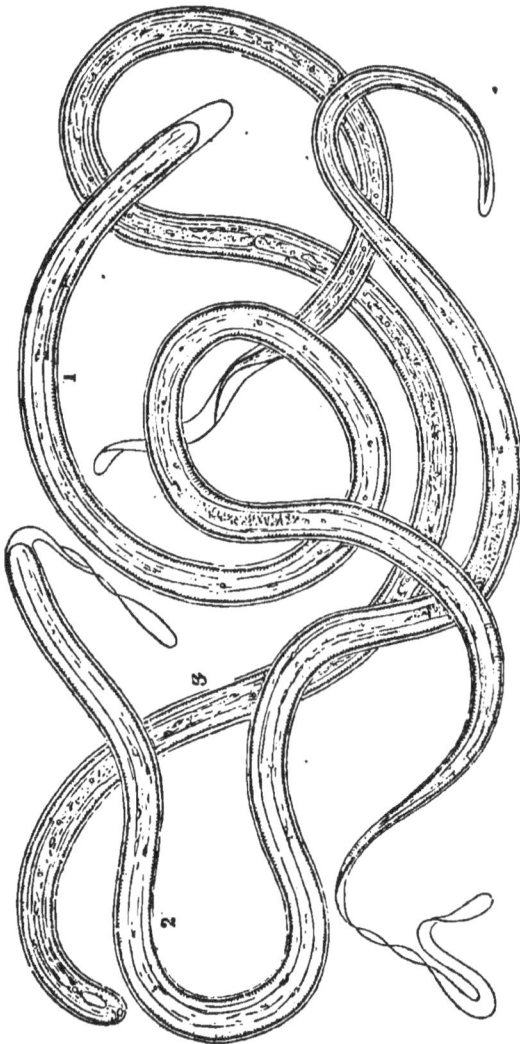

VARIOUS APPEARANCES PRESENTED BY A SINGLE HÆMATOZOON,
AS OBSERVED UNDER ⅓' AND ¹⁄₁₂' IMMERSION OBJECT GLASSES.

Fig. III.

As the movements of the Filaria became a little

The whip-like ante-
rior and posterior pro-
longations:

slower, it was seen that the striæ
were not on its outer coat, but con-
fined to the body of the worm, and that the tail, which

almost always under the $\frac{1}{4}''$ objective looked like a lash, was not so in reality, but that, every now and then, it could be seen flapping against the corpuscles like a fin—sometimes vertically, sometimes horizontally, and then becoming folded upon itself like a ribbon (Fig. III, 1), a condition which I had already observed and figured two-and-a-half years ago without knowing what it was. Precisely similar phenomena were observed to occur at the opposite terminal extremity (Fig. III, 2).

It was, however, only after the lapse of fully five hours' careful watching, the activity of the Hæmatozoon having considerably subsided, that the real nature of what appeared to be the rapid protrusion and retraction of the delicate membrane at the oral and caudal terminations, was discovered. An unusually long tail was seen to be trailing after the 'body' of the Filaria for several seconds, and whilst thus being dragged, fortunately, it remained exactly in focus, when suddenly the ribbon-like-folds were straightened by the darting of the pointed extremity of the worm into the very tip of this hyaline filament (Fig. III, 2). Scarcely had this taken place than the tail was again retracted and the ribbon-like appendage became evident once more; whereupon the ribbon-like filament at the other extremity was suddenly straightened in a similar manner and the 'head' rapidly projected into the very tip.

The explanation of the appearances presented by the filaments.

The Hæmatozoon may, therefore, be said to be *enveloped in an extremely delicate tube, closed at both ends*, within which it is capable of elongating or short-ening itself. This tube, like the sarcolemma of muscu-lar fibres, is without any visible structure, is perfectly transparent, and, but for the difference between it and the fluid in which it is immersed in its power of refract-ing light, which allows of its margins or folds being brought into view, it could not be demonstrated.

The fact of its being thus enclosed seems to show that in the present stage of its exist-ence, the 'home' of this Filaria is in the blood; it has no visible means of perforating the tissues; moreover, although constantly observed to be in a state of great activity, it does not seem to manifest any special tendency to migration, and is apparently dependent on the current of the blood for its transference from place to place; its movements, therefore, within this enveloping tube, appear to be as limited as those of any other animal enclosed within a sac.

Probability of the blood being the 'home' of the Filaria in the present stage of its development.

As has been already stated, a short chain of aggregated molecules, probably repre-senting the rudiments of an intes-tinal canal, is frequently seen towards the centre (Fig. III, 1), but the rest of the entire length is at first uninterruptedly clear, although not transparent. But during the time the details described in the preceding paragraphs were observed, and they became more and more evident as the activity of the Hæmatozoon diminished, the appear-

Further anatomical details suggestive of the existence of an oral aperture, and of the development of a di-gestive canal.

ance throughout became granular or rather molecular. A bright spot also became very evident at the terminal point of the anterior portion, which, as already remarked, is extremely suggestive of an oral aperture, and immediately below this a somewhat elongated vacuole. From this downwards, until about the junction of the middle with the lower third, or perhaps a little nearer the middle, a more or less clearly differentiated œsophagus (?) became likewise discernible, and appeared to have a cœcal termination, but beyond this, until the caudal extremity was reached, the continuation of the digestive tract was less clearly defined (Fig. III, 3).*

It then became too dark to continue the observation, and by the next morning the Filaria had become uniformly molecular, all appearances suggestive of internal organs having vanished, although it still continued to coil itself languidly amongst the blood corpuscles.

Such is the minute anatomy of the Hæmatozoon as far as I have been able to make it out. What has here been recorded has now been repeatedly observed, and may be observed by any one possessing a good $\frac{1}{8}''$ or $\frac{1}{12}''$ immersion lens, and a microscope provided with good arrangements for illumination. The simple detection, however, of the Hæmatozoa, when present in the blood, is simply a question of patience, and not dependent

---

* For the care with which these wood-cuts have been executed, I am much indebted to H. H. Locke. Esq., Principal of the Government School of Art, under whose superintendence they were engraved.

on any special perfection in the magnifying powers employed.

The average diameter of the Hæmatozoon, as usu-

Measurements. ally found, is, as already stated, about that of a red-blood corpuscle, and its average length about 46 times that of its greatest width; that is to say, its greatest transverse diameter is about $\frac{1}{3500}$ of an inch and its length about $\frac{1}{75}$th of an inch. These are about the measurements most frequently met with, but I have occasionally seen specimens not more than half this size. The largest specimen which I have measured was found to be slightly over $\frac{1}{3000}$th of an inch in width and about $\frac{1}{8}$th of an inch in length, whereas the smallest was only $\frac{1}{7000}$th of an inch in width and $\frac{1}{125}$th of an inch in length: the relative proportion between the length and the greatest width being as 1 to 45 in the largest and 1 to 56 in the smallest; the width, therefore, gaining somewhat in proportion to the length as the total dimensions increase.

From what has been above stated concerning the power of extension and contraction possessed by the Hæmatozoon, it will be perceived that these measurements are subject to variations during life; and, as death may occur when the Filaria may happen to be in either of these conditions, the relative proportion between the length and the breadth may then also be found to vary somewhat.

In order to prevent misconception, it may perhaps be well to compare these measurements with those of two well-known Nematode helminths which are occasionally found in the tissues of the human body, *viz.*, the Muscle-trichina and the Guinea-worm, or rather its contained embryos. Both of these parasites present a certain degree of likeness to the Filaria described in this paper. The first named is found in the muscular tissue; the second in the cellular tissue; and the third in the blood. All three present transverse markings, more or less evident—in the Guinea-worm embryo they are particularly distinct; but beyond these features the similarity between them appears to cease.

Comparisons instituted between the Hæmatozoon, the Trichina spiralis and the Guinea-worm.

They differ from each other in size, in form, and in the relative proportions of length to breadth—setting aside altogether the great difference which exists between their minute internal organisation.

As to size and form the Hæmatozoon approximates more closely to the Filaria medinensis or Guinea-worm embryo, than to the larval stage of the Trichina spiralis, though much smaller than either, especially in breadth. The average length of samples of the former which I possess is $\frac{1}{22}$nd of an inch, and the breadth $\frac{1}{1000}''$, so that the breadth to the length is as 1 to 31: whereas the specimens of Trichina with which these comparisons were made averaged $\frac{1}{25}$th of an inch in length and $\frac{1}{700}$th in width; so that they are only 28 times the length of

Differences of size and of the relative proportion of breadth to length:

their greatest transvetse diameter. It will be remembered that these proportions in the case of the Hæmatozoon have been referred to as being on the average 1 to 46.

A still greater dissimilarity between these helminths than the disparity in size and relative proportions, is the totally different aspect presented by their anterior and

Different aspects presented by the anterior and posterior terminations of all three.

posterior extremities; this is sufficiently evident without referring to the minute structural anatomy of the parts. The cephalic end of the Trichina is almost pointed and its caudal termination blunt; whereas, although the anterior extremity of the two Filariæ agrees in the matter of being somewhat rounded and the posterior end in both comes to a very fine point, nevertheless, the relative proportion between the tail of the one and that of the other is sufficiently great to present a marked difference—the tail of the Dracunculus being nearly $\frac{1}{3}$rd, whereas that of the Hæmatozoon is not, at the utmost, more than $\frac{1}{8}$th of the entire length. Of course this is exclusive of the hyaline tube within which the latter is enclosed. Possibly when live young Dracunculi shall have been as carefully examined and described as the lifeless specimens have been by Mr. Busk and Dr. Bastian, the similarity between the Filariæ will become more evident.*

---

* Since this paragraph was in type, I have, however, had ample opportunity of satisfying myself on this matter by the examination of numerous young Dracunculi in all stages of development obtained from a patient admitted into the General Hospital under the care of Dr. Coull Mackenzie, suffering from Guinea-worm, but I find that there is even less resemblance between the Filariæ during life than was suggested by lifeless specimens.

The comparisons just instituted between the three helminths referred to will, perhaps, be more clearly understood by throwing these details into a tabulated form* :—

	Average breadth.	Average length.	Relative proportion of breadth to length	Aspect presented by		Relative length of tail to total length.
				Head.	Tail.	
Trichina (of muscle)	$\frac{1}{700}''$	$\frac{1}{25}''$	1 to 28	Pointed	Blunt ...	......
Dracunculus(embryo)	$\frac{1}{1000}''$	$\frac{1}{32}''$	1 to 31	Rounded	Acutely pointed.	1 to 3½
Human Hæmatozoon	$\frac{1}{3500}''$	$\frac{1}{75}''$	1 to 46	Ditto ...	Ditto	1 to 8

The part which the Hæmatozoon appears to take in the production of disease will become still more evident when the condition of the kidneys and suprarenal capsules, referred to in a previous paragraph (page 17) as having been obtained from a patient who died of Chyluria, has been described.

To the naked eye none of these organs presented any marked deviation from the normal standard, except that the kidneys were more than usually lobulated, and, that on section several of the pyramids, especially near their apices, presented a

Tallowy appearance presented by some of the pyramids of the kidney.

smooth, tallowy appearance, suggestive of amyloid disease. No approach to the characteristic iodine re-action could, however,

---

* The measurements here introduced of the young Trichinæ and Dracunculi do not materially differ from those generally given. For the sake of uniformity, however, it was considered advisable to measure all three with the same micrometer-scale.

be obtained; but when longitudinal sections were subjected to microscopic examination, numerous translucent
oil-like tubules of a somewhat varicose appearance
could be observed running alongside the uriniferous
tubes as if the lymphatics or minute blood vessels of
the part had become plugged. These sections when
placed in boiling ether, and afterwards subjected to prolonged maceration in it, did not appear to be materially
affected by the process—the translucent oil-like tubules
being quite as evident as before.

No other morbid changes could be detected as
having taken place in either the
tubular or cortical tissue of the
kidneys, but in every fragment, no
matter from what part of the kidneys removed, numerous microscopic Filariæ were invariably obtained, if
the tissue had been properly teased, precisely analogous
to those which had been detected in the blood and
in the urine during life. Teased fragments of the supra-
renal capsules yielded similar specimens. On slitting
open any portion of the renal artery, from its entrance
into the kidney as far inwards as I was able to follow
its ramifications, and gently scraping its inner surface
with a scalpel, numerous Hæmatozoa could always be
obtained. The renal vein when similarly examined also
yielded specimens of the Filariæ, but they did not seem
to be so numerous in it.

The vessels themselves did not appear to be
diseased, and such of the branches as could be seen
with the naked eye did not strike me as being

The kidneys and supra-renal capsules contained numerous Hæmatozoa.

abnormally large. But whether the microscopic ramifications and the capillaries were distended or otherwise (in the absence of properly injected preparations of the organs) could not well be ascertained.

Having traced the course of the Hæmatozoon from the blood through its channel into the urine, the peculiar appearances presented by this secretion will now be very briefly considered.

The chemical constitution of chylous or milky-urine is so well known that it is not necessary to do more than refer to the principal features which it presents. It is, as the term applied to it conveys, more or less perfectly white, has a faint odour of milk, which is heightened by warmth ; and, like that secretion, may be passed through several layers of filtering paper without materially modifying its colour. Usually it is of low specific gravity—from 1006 to 1018—and manifests a slightly acid re-action to test paper. As a rule, the more it approaches the appearance of milk, the more readily and firmly does coagulation take place. When the presence of blood is a prominent feature in it, curdling takes place still more perfectly, but the early addition of solutions of ammonia, sulphate of soda, or nitrate of potash retards, if it does not completely prevent this change ; frequently, however, the process has already commenced before the escape of the fluid from the bladder.

The general appearance presented by chylous urine.

The elaborate analyses which have from time to time been made of the urine in this condition, as well as such simple analyses as I have been able to conduct, have not tended to show

Its chemical constituents ;

that there is any organic or inorganic substance in the
secretion, but what already exists in the nutritive fluids
of the body, or that any new *chemical* combination
has been called into existence. With regard to the
alleged presence of sugar in this kind of urine, my
attempts to detect it have been entirely negative.

In short, the urine appears merely to deviate from
In what way' it dif- the healthy standard in so far that
fers chemically from
healthy urine. it contains an abnormal amount of
fatty and fibro-albuminous material, with, perhaps, a
diminution in the percentage of urea; in connection
with this, however, I may add that in the cases noted
which presented a low specific gravity (1006—1010) the
quantity voided had been considerable, from 60 to 70
ounces in the course of the 24 hours.

On no occasion have I been able to detect 'casts'
of renal tubules in urine of this nature, even in cases
where previous attacks of the malady had occurred.

When subjected to microscopic examination, this
Microscopic examina- kind of urine presents a finely mole-
tion of chylous urine;
cular appearance; when recent, how-
ever, scarcely any distinct oil-globules, such as are constantly observed in milk, are present; but when acetic
acid is added, followed by a little warm ether, this
'molecular base' becomes replaced by large globules of
fat, which may be seen to form whilst the re-agents
are being applied. In the meshes of the coagulated
substance numerous granular cells are seen, apparently
identical with those of chyle, lymph, or the white cells
of the blood; and, generally, a sprinkling, more or less
marked, of red-blood corpuscles.

Besides these, if the shred of coagulum on the
slide has been properly 'teased,' the
Filariæ, described in the preceding
pages, will also be usually found. As before stated, in
not a single case which has come under my notice have
they been absent. However, they *may* not be present in
every sample of Chylous urine examined, or rather I
should say, the numbers present may be so few as to
elude detection.

The detection of the Filariæ sometimes very difficult;

With regard to the size of the Filariæ which are
met with in the urine, it may be
observed that they present the same
measurements as those met with in the blood; some
of the largest as well as some of the smallest examples
have been found in this secretion.

Dimensions of the Filariæ found in it.

The importance of bearing in mind the difficulty
that is sometimes experienced in dis-
covering the Filaria in the urine also,
may possibly be more strongly im-
pressed by the narration of an illustrative case, which
will, moreover, serve to draw attention to other impor-
tant matters bearing on the question of infection with
Hæmatozoa :—

A case illustrative of the difficulty sometimes experienced in detecting Filariæ in the urine.

A European, age 38, formerly in the army, was
kindly sent to me by Dr. McConnell, with a note stating
that the man was amongst his out-patients and had
been suffering, and was even suffering a little still, from
Chyluria. The medical history which I gathered from
the man was, briefly told, as follows: Has suffered more
or less constantly for five years from what he believes
to be 'chronic dysentery.' This came on during his

residence in Mysore. Eight months after the advent of
the intestinal affection, he observed that the urine passed
towards the middle of the day was white, but was not
so in the early morning. His hearing and sight became
affected about the same time, and have remained imper-
fect since, although there is nothing to be observed
wrong about either set of organs.

The urine at the time he paid me a visit did not
seem to be particularly affected, merely a little cloudy, but was albuminous. A little carbolic acid solution
having been added to it, it was set aside in a conical
vessel, and subsequently the sediment removed by means
of a pipette for microscopic examination. This is
usually the method adopted by me in cases when the
fluid does not coagulate, or when, after coagulation has
taken place, it has become liquified.

Mode of preserving Chylous urine for subsequent microscopic examination.

Slide after slide was examined in vain, still I felt
so convinced that, as there had been
a distinct history of Chyluria, and
as the urine was still albuminous, but contained no
' casts,' the original cause had not entirely disappeared.
Eventually after searching for about four hours, three
excellent specimens of the Filariæ were obtained, one
of which I forthwith despatched to Dr. McConnell.
A week afterwards the patient returned, but I failed to
detect a single worm in the specimen of urine which
was obtained on this occasion. He came a third time,
after an interval of about another week, when Filariæ
were detected in the sediment without much delay.

Eventual result of numerous examinations.

Several preparations of the blood were also examined, but the Hæmatozoa were not detected in this fluid. Were, however, a couple of *ounces* of the blood examined (coagulation being prevented by the addition of a neutral salt), instead of a couple of drops, doubtless the sediment would contain plenty of the Filariæ, seeing that a few must have actually passed out of it through the kidneys, as we have already seen that they are not localised in these organs, the latter simply acting as one of the channels through which they may escape out of the circulation.

The phenomena associated with Chyluria are so well known that it is not deemed necessary to give more than the salient features of the malady, more especially such, as are exemplified in the cases referred to in this paper; which, in the main, correspond very closely with the cases that have been from time to time recorded by others.

The salient features of Chyluria:—

In the first place it is to be noted that the malady is decidedly *localised* as to its origin. As far as I have been able to ascertain, the only cases on record have occurred in persons who have at some period of their lives inhabited the East or West Indies, some parts of Africa, Bermuda, Brazil or the Mauritius; so that all writers agree, no matter to what particular cause the disease has been referred, that it is intimately related to a tropical climate. Simple removal, however, from such climate has not sufficed to prevent a recurrence of the disease in England or in other parts of Europe.

Originates in certain countries;

F

Secondly, it is noticeable, that the disease, as ma-
nifested by the milky appearance of

Suddenness of attack.

the urine, comes on very *suddenly*,
not only on the first, but on succeeding occasions
also; this peculiarity to my mind points to a local cause
in the system, rather than to a generally distributed
functional disorder.

Thirdly, there is a complete absence of casts of the
tubules of the kidney in the urine,

No signs of organic kidney disease.

notwithstanding the large amount of
albuminoid elements present.

And, fourthly, it is frequently associated with more
or less distinctly marked symptoms of

Concomitant maladies.

various other obscure diseases, such as
partial deafness; diarrhœa, often very persistent; chronic
conjunctivitis or some more deeply-seated defect in the
visual organs; and sometimes temporary swellings of
the face or extremities.

These varied complications may, I believe, be
very satisfactorily accounted for now

The probable cause and pathology of chyluria, &c.

that it has been ascertained that the
nutritive channels of the tissues, even to their most
minute ramifications, are inhabited by numberless living
Hæmatozoa, which, accidentally or otherwise accumulat-
ing in any particular set of these channels, may lead to
local stoppages in the flow of the nutritive fluids and to
rupture of the extremely delicate walls of the capillaries,
lacteals or lymphatics. The extreme activity of the
Filariæ, especially should a bundle of them accumulate
in one particular spot, would doubtless materially and in-

giving rise to rupture—for, as is well known, the walls of these channels are extremely delicate, those of the lymphatic system being especially so. The resulting phenomena, such as the escape of the nutritive fluid and of the Filariæ contained within the ruptured channel into the excretory ducts belonging to the part, appears to me to be so simple a procedure that to dilate on its mechanism would be quite superfluous. When the fissure becomes plugged or healed, these unusual symptoms naturally disappear, induced as they are due to some changes induced in the former, or in the fluid in contact with the nerve

It would seem, therefore, that the milky appearance of the urine is merely one of the symptoms of the existence of this Filaria in the nutritive channels of the body.

It must not, however, be inferred that it would refer all cases of Chyluria to this cause. Doubtless, the combination of various other circumstances might produce similar phenomena, just as various obstructing causes, such as the pressure of tumours, diseased condition of the vessels, &c., may produce the exudation of milky fluid in various parts of the body—from the abdominal walls, the groin, the axilla, the thigh and other parts, such as are constantly being reported in medical journals. Nevertheless, cases occurring in warm countries, or in persons who had formerly resided in them, appear to indicate that the disease is not dependent on such mechanical or pathological causes as these.

The same remarks apply to the etiology of the

Cause of various con-
ditions which may or
may not be associated
with Chyluria :—
various other phenomena enumerated as the more or less frequent concomitants of Chyluria, without much modification—for even should actual rupture not take place, local congestions may be induced or very trifling fissures, which might yet be sufficient to interfere with the functions of delicate organs, or it may be, as in the case of the eye or ear, that the mischief may be chiefly due to some changes induced in the refractive media of the former, or in the fluid in contact with the nerve filaments in the latter.

The intestinal affection, if in reality connected with

The intestinal affec-
tion, &c.;
the entrance or exit of these Filariæ, deserves special attention. The only *known* symptom from which the man in whose blood this helminth was first discovered, was severe diarrhœa; the commencement of the illness of another man is dated from a similar attack, which developed itself into what is described as " chronic dysentery," on which the usual medicines appear to have had no influence, for during the last five years the disease came and went without reference to medical treatment—the chylous condition of the urine having been equally irregular in its appearance and disappearance. It will be remembered that the intestinal affection commenced eight months before the urinary symptoms appeared; moreover, in the woman whose autopsy has been recorded on a previous page, ' tubercular-like' ulcers were found in the intestines, as also in the lungs. All these occurrences, especially

when taken in connection with what is *known* to occur in connection with the migration of several parasites, are too prominently associated with the history of the cases in which these Hæmatozoa were detected to permit of the subject being passed over without comment.

With reference to the 'granular-lid' condition of one of the patients affected with Hæmatozoa, it has been demonstrated since the earlier pages of this paper was in type, that, not only had congestion resulted from the presence of the Filariæ, but actual rupture and escape of one of them at least occurred, either through the channel of the lachrymal or of a Meibomian gland (see foot-note, page 13).

Some forms of 'granular-lids';

Although feeling convinced that Chyluria and other morbid phenomena are induced by the presence of these microscopic Filariæ in the circulation, still, unless it can be shown that they may have a prolonged existence in this condition, it will be difficult to reconcile this opinion with the fact that the malady so frequently recurs in the same individual. It seems unlikely that the same person should become re-infected with Hæmatozoa several times, and especially that re-infection should occur after years of residence in England, where probably this particular Filaria is not indigenous.

The recurring attacks of Chyluria.

Not having been able to watch the progress of isolated cases for a sufficiently long period to judge whether or not all the Hæmatozoa in the system escape during a single attack of Chyluria (the period of their existence in the

Hæmatozoa in lower animals—in dogs, &c.

stage in which they are found in the blood (having ex-
pired), nor having succeeded in prolonging their existence by artificial cultivation, in serum, moist sand or
saliva, beyond a period of three days, it will be necessary
for me to refer briefly to a few of the recorded instances of Hæmatozoa occurring in lower animals, so
as to fill up the gap in the chain of evidence as far as
possible.

Foremost among the recorded particulars concerning
Hæmatozoa are those of MM. Grube and Delafond,
which were presented in their 'Mémoire' to the French
Academy of Science.[*] These gentlemen, during a period
of nine years, made observations on 29 dogs in whose
blood on an average 55,000 microscopic worms were
estimated to exist. The diameter of these was somewhat less than that of a red-blood corpuscle; the length,
however, is not given in this communication, but I find
this referred to in one of the early volumes of the
'Lancet' (1843) as being about ½₄th of an inch, which
is somewhat smaller than other human Hæmatozoon.
These dogs were under observation for periods varying
from several months to five years, during which the
state of the blood remained unchanged. Post-mortem
examinations appear to have been conducted with great
care at all seasons of the year, but on one occasion only
were what the authors deemed to be the parent-worms
discovered. Six of these were found to be lodged in a
large recently-formed clot in the right ventricle—four
being females and two males. The size of these was by

---

[*] Mémoire sur le ver filaire qui vit dans le sang du chien domestique 'Comptes Rendus' T. XXXIV, p. 9.

no means microscopic, being from 5 to 7 inches in length (14 to 20 centimètres)*, and from $\frac{1}{25}$th to $\frac{1}{17}$th of an inch transversely. Schneider questions whether these were the parent-worms of the microscopic Filariæ;† others state that they had simply found their way to the heart from the intestines by accident, because this observation of MM. Grube and Delafond, although published about twenty years, has never been confirmed. Leuckart, who, however, expresses no opinion on this particular subject, refers to these observations as an illustration of the fact that, with the exception of the Trichina spiralis, not a single nematode has been observed to infect its own bearer—the Hæmatozoa of dogs as well as of frogs never having been observed to develop into mature helminths as long as they remained in the blood.

In a highly interesting paper read by Dr. Cobbold before the Linnean Society in 1867,§ it is more than hinted at that the Hæmatozoa referred to by MM. Grube and Delafond were the brood of 'Spiroptera sanguinolenta,' so commonly found in the heart of dogs in China, but nothing is mentioned concerning the microscopic examination of the blood of these animals. In a foot note it is stated that Dr. Lamprey had forwarded specimens to the Netley Museum. Should these be still in a good state of preservation, it would be a great matter if Professor Aitken would re-examine

* Not all erroneously stated in some English Works on Helminthology from one-half to three-fourths of an inch.
† 'Monographie der Nematoden', 1866, p. 88.
‡ 'Menschlichen Parasiten', Vol. II. Part 1, p. 102.
§ 'Journal of Linn. Soc.—Zoology', Vol. IX.

the specimens, especially as to the minute structure of the contained embryos, if there be any, and publish the result.

As regards the blood and heart of dogs in India, out of nearly 200 dogs examined by Dr. D. D. Cunningham and myself, in connection with various experiments, in no instance were any such helminths detected, so that the canine Filariæ of France and China would appear not to be found in Bengal.

Dogs in Bengal not affected with Hæmatozoa.

Dr. G. E. Dobson has drawn my attention to a description of mature Filariæ found by M. Joly in the heart of a seal; the female worm is stated to have been stuffed throughout its entire length with ova and embryos; the latter measured $\frac{1}{40}$th to $\frac{1}{36}$th of an inch in length and $\frac{1}{2500}$th in breadth, but the author does not consider that they could circulate with the blood through the capillary vessels.*

Such, in a few words, is the present state of our knowledge of the principal Hæmatozoa affecting lower animals; and from these records alone would our inferences have had to be made in regard to the particular question as to the possible duration of the Human Hæmatozoon, were it not for a rather strange coincidence.

The foregoing account had just been transcribed from my notes, when I had occasion to visit the Government Printing Establishment, where, to my utter surprise, I saw, busily putting into type a portion of the

The Filariæ detected in the blood of a man in whose urine they existed more than 2½ years previously.

---

* Ann. Mag. Nat. Hist., 1858, p. 400.

foregoing pages, the very man in whose urine these Filariæ were first detected—more than two and half years ago ! Being rather below the average in intelligence, he had not the remotest idea to what the manuscript referred. *

At my request he called upon me in the afternoon, and I learnt from him then, that his urine had been perfectly healthy ever since he left the hospital, about April 1870,—it certainly looks healthy now, and is quite free from albument. I prepared seven slides from blood obtained by pricking the middle finger of one hand, and three slides from the same finger of the other hand. On seeing me do this, the man enquired why I had made so many preparations, as on a former occasion I had only taken *one* slide; a circumstance, by the way, which I had quite forgotten; certainly I had not discovered Hæmatozoa. This little incident also conveys its lesson; had I taken a dozen slides on the first occasion instead of one, the date of the detection of the Filaria in the blood would probably have been simultaneous with their detection in the urine. In the first four or five preparations examined nothing could be observed; in the two next taken up, one belonging to each hand, Hæmatozoa were detected, very active, but in no way differing from the excellent live specimens which I had obtained in his urine long ago, and in no way differing from the

---

* See page 5—6.

† On referring to my notes of this case, I find that, at the time when he left the Hospital, the albumen had disappeared from his urine, and that Filariæ could no longer be detected in it.

Filariæ since detected in the urine and blood of so many persons. The measurements of two specimens were taken on the following morning after their activity had subsided; one was $\frac{1}{76}$th of an inch in length by $\frac{1}{3500}$th in breadth, and the other $\frac{1}{80}$th by $\frac{1}{3500}$th.

Here is, therefore, definite information more satisfactory than that to be obtained by instituting comparisons between the Hæmatozoa of man and of animals,

The prolonged existence of Hæmatozoa sufficient to account for recurrency of the malady.

that, not only may those found in man *live for a period of more than* 2½ *years, for certain,* but that there is no evidence that they have any tendency to develop beyond a certain stage, so long as they remain in the circulation. For aught we know to the contrary, these Filariæ may live for many years, and thus, at any moment, no matter how long after a previous attack, nor in what country the person may reside, he may be surprised by the sudden accession of Chyluria or any other obscure disease, such as will readily be understood by the physician when he becomes aware of the state of the blood.

If after the first brood of young Filariæ, there be no provision for other broods to follow, then every attack would be a step towards permanent recovery, but of this I know nothing at present, although some of the cases recorded appear to warrant such an inference. Nor have I any definite knowledge as to how the blood originally becomes infected; to hint that it is possible if not probable that the Filaria may eventually be traced to the tank—either to its water or its fish, is the utmost that can be done.

Many other interesting questions suggest themselves as matters for future enquiry; such for example
as to whether the fœtus *in utero* is infected by the
mother's blood: cases have been recorded which seem
to favour such a supposition; such instances, however,
may have been due to the particular localities in which
the persons resided—parents and children having for
generations been subject alike to the same influences.
On one occasion, I attempted to solve this question, but
the mother, who herself was very averse to having her
finger pricked, peremptorily refused to submit the child
to a similar trivial operation.

This paper having considerably exceeded the limits
originally intended, it may be that
A summary of the the leading facts referred to have
principal facts and
inferences recorded. become obscured by the digressions that have been
necessarily made, so that, before concluding, a short
summary of the observations and of the inferences
which have been deduced therefrom may be advan-
tageous :—

(1). The blood of persons who have lived in a
tropical country is occasionally invaded by living
microscopic Filariæ, hitherto not identified with any
known species, which may continue in the system for
months or years without any marked evil consequences
being observed; but which may, on the contrary, give
rise to serious disease, and ultimately be the cause of
death :

(2). The phenomena which may be induced by the
blood being thus affected are probably due to the me-

chanical interruption offered (by the accidental aggregation, perhaps, of the Hæmatozoa), to the flow of the
nutritive fluids of the body in various channels, giving
rise to the obstruction of the current within them, or
to rupture of their extremely delicate walls, and thus
causing the contents of the lacteals, lymphatics or capillaries, to escape into the most convenient excretory
channel. Such escaped fluid, as has been demonstrated
in the case of the urinary and lachrymal or Meibomian
secretion, may be the means of carrying some of the
Filariæ with it out of the circulation. These occurrences are liable to return after long intervals—so long
in fact as the Filariæ continue to dwell in the blood :

(3). As a rule, a Chylous condition of the urine
is only one of the *symptoms* of this state of the circulation, although it appears to be the most characteristic
symptom which we are at present aware of :

(4). And, lastly, it appears probable that some of
the hitherto inexplicable phenomena, by which certain
tropical diseases are characterised, may eventually be
traced to the same, or to an allied condition.

The importance of a careful microscopical examination of the blood of persons suffering from obscure
diseases, in tropical countries especially, is therefore more
than ever evident, and opens up a new and most im-
portant field of enquiry—referring as it does to a hitherto unknown diseased condition.

Calcutta, October 1872.